2/21/00

ISBN 1 85854 722 9
Published by Brimax Books Ltd, Newmarket,
England, CB8 7AU, 1999.
Printed in France.

Little Lamb

By Gill Davies

Illustrated by Sally Hynard

Today is cold and wintry;
Snowflakes drift and float.
Father comes into the kitchen –
Something's tucked inside his coat.

He says, "Here is a little lamb.
He was only born today.
He needs some milk to drink
And somewhere warm to play."

Little Lamb is very small.

He wobbles when he stands.

His wool feels soft and tickly.

He baas and butts our hands.

Mother gives the lamb some milk.
He sucks the bottle greedily.
Then he sighs with happiness
And rests his head on Mother's knee.

Little Lamb soon grows much bigger.
Now he can go outside again.
He wanders round the farmyard
Where the ducks quack in the rain.

He plays with all the other lambs,
Skipping in the sun.
But when I say, "Hi, Little Lamb!"
Up to me he'll run.

I miss him now he's gone outside
So I talk to him out there.
I tell him what's been happening
And stroke his bubbly hair.

He nibbles at my fingers,
Then he bounces off to play.
I am so very glad
That with us he came to stay.